I0822914

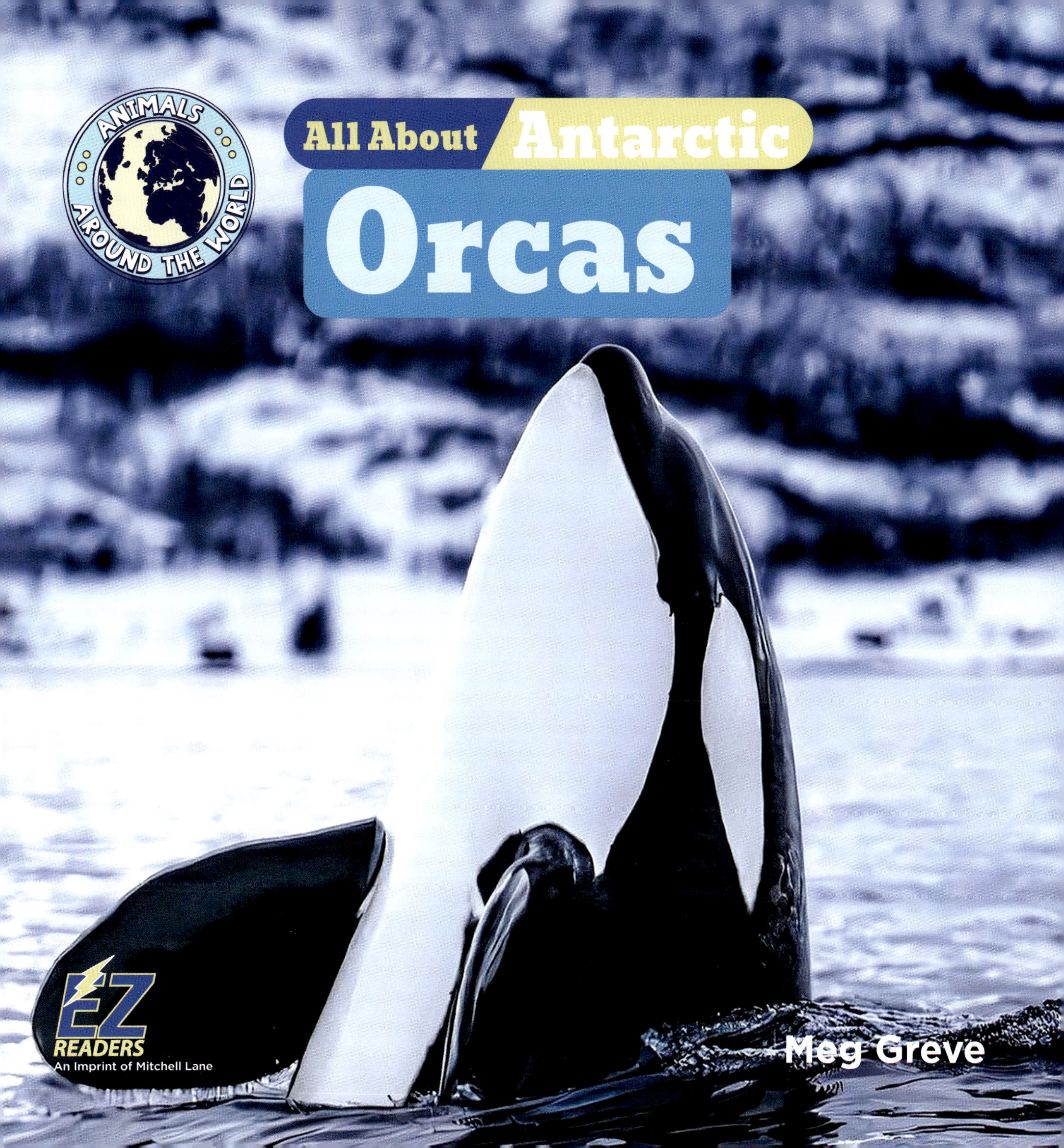
Animals Around the World
All About Antarctic
Orcas
EZ Readers
An Imprint of Mitchell Lane
Meg Greve

Creating Young Nonfiction Readers

EZ Readers lets children delve into nonfiction at beginning reading levels. Young readers are introduced to new concepts, facts, ideas, and vocabulary.

Tips for Reading Nonfiction with Beginning Readers

Talk about Nonfiction

Begin by explaining that nonfiction books give us information that is true. The book will be organized around a specific topic or idea, and we may learn new facts through reading.

Look at the Parts

Most nonfiction books have helpful features. Our *EZ Readers* include color photographs and graphic aids, a table of contents, a glossary, and an index. Share the purpose of these features with your reader.

Color Photos and Graphic Aids

A lot of information can be found by "reading" photos, charts, maps, and other graphic aids found within nonfiction texts. Help your reader learn more about the different ways information can be displayed.

Table of Contents

Located at the front of the book, this list shows the big ideas within the text and the page numbers where they can be found.

Glossary

Located at the back of the book, the glossary defines key words and phrases that are related to the topic. These words and phrases can be found in the text in colored type.

Index

Located at the back of the book, an index is an alphabetical list of topics and the page numbers where they can be found.

With a little help and guidance about reading nonfiction, you can feel good about introducing a young reader to the world of *EZ Readers* nonfiction books.

Mitchell Lane
PUBLISHERS

2001 SW 31st Avenue
Hallandale, FL 33009
mitchelllanepub.com

First Edition, 2026.

Author: Meg Greve
Designer: Rhea Magaro
Editor: Kim Thompson

Names/credits:
Title: All about Antarctic Orcas / by Meg Greve
Description: Hallandale, FL :
Mitchell Lane Publishers, [2026]

Series: Animals Around the World
Library bound ISBN: 979-8-89260-518-2
eBook ISBN: 979-8-89260-523-6

Library of Congress Control Number: 2025933682

EZ Readers is an imprint of
Mitchell Lane Publishers

PHOTO CREDITS
Shutterstock: Dreamworld Creations, cover, 1; MuhammadHanif1, 5; elena_prosvirova, 8; Foto4440, 10; longtaildog, 12; Guillermo El Oso, 14; robert mcgillivry, 16; deer boy, 22; Alamy: Wildest Animal, 7; Dreamstime: slowmotiongli, 18;

Contents

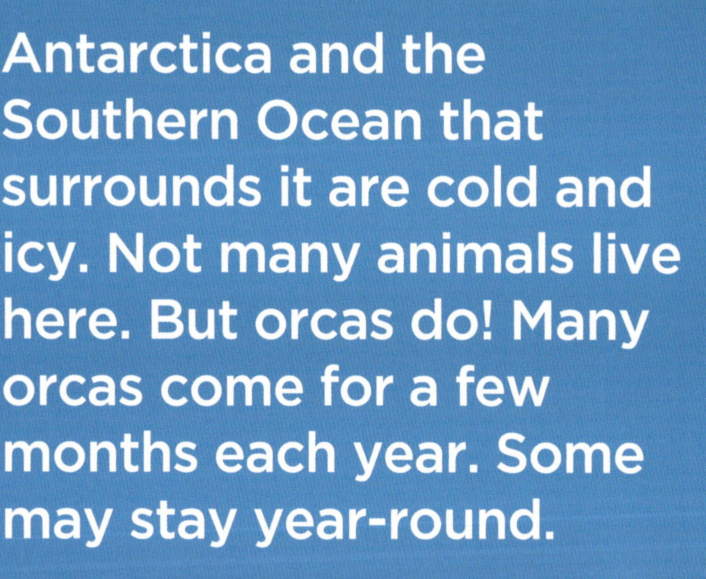

Antarctica and the Southern Ocean that surrounds it are cold and icy. Not many animals live here. But orcas do! Many orcas come for a few months each year. Some may stay year-round.

Orcas are **marine mammals**. They are also called killer whales. They can be 32 feet (10 meters) long. The longest orcas are about the size of a train car!

Orcas have black backs. They have white bellies. Looked at from above, orcas blend into the dark ocean depths. Looked at from below, they blend into sunlight at the ocean's surface.

An orca's **dorsal** fin keeps it from tipping over. Each dorsal fin has a different shape and pattern, like a fingerprint. Side fins are like arms. They help orcas steer through the water and make quick turns.

An orca's strong tail has two sides called **flukes**. Orcas can "stand" on their tails. They can poke their heads above water to look around. This is called spy-hopping.

Orcas do not chew. Instead, they use sharp teeth to grasp **prey** and tear it into chunks. They catch squid, fish, penguins, seals, and smaller whales. They have no **predators**.

Orcas live in groups called pods. There are up to 50 orcas in a pod. Pods hunt together. They may help each other splash a seal off the ice. They may surround fish and hit them with their flukes.

Orcas have one **calf** about every five years. The baby drinks milk from its mother's body for the first year. The mother feeds it until it is old enough to hunt.

Babies learn everything from their mothers. They learn to talk using clicks and squeaks. When babies grow up, they stay with their pods. Orcas can live for 90 years!

Where Do Orcas Live?

Orcas live in oceans around the world. They may move from place to place in different seasons. Many orcas visit Antarctica or stay there year-round. The best times to see orcas in Antarctica are February and March.

Interesting Facts

- Only half of an orca's brain sleeps at one time. The other half stays awake to help the animal return to the ocean surface to breathe.
- Orcas share almost all their meals with members of their pod.
- Under their skin, orcas have a thick layer of blubber that keeps them warm in cold water.
- Orcas sometimes smack their flukes on the surface of the water. This is called a tail lob.

Parts of an Orca

blowhole
The blowhole is used for breathing. Orcas do not breathe through their mouths.

dorsal fin
The dorsal fin helps an orca stay up. It also helps the animal stay at a comfortable temperature.

eye patch
There is a patch of white behind each eye. No two patches are the same.

eyes
Orcas have great eyesight. This helps them spot prey underwater.

fluke
Each fluke is one half of an orca's double-sided tail.

melon
The melon is a fatty spot on the orca's head. It helps orcas hear and communicate.

pectoral fins
Side fins help orcas swim and steer.

Glossary

calf (kaf)
A baby whale

dorsal (DOR-suhl)
Located on the back

flukes (flooks)
The left and right lobes of an orca's tail

marine mammals (muh-REEN MAM-uhlz)
Warm-blooded animals that spend all or most of their lives in the ocean, that have backbones and fur or hair, that give birth to live babies, and that feed their babies milk from the mother

predators (PRED-uh-turz)
Animals that hunt other animals for food

prey (pray)
Animals that are hunted by other animals for food

Index

Further Reading

Bader, Bonnie. *My Little Golden Book about Whales.* Random House USA, 2024.

Chanez, Katie. *Killer Whales.* Jump!, 2023.

On the Internet

Active Wild: Antarctica Facts for Kids
www.activewild.com/antarctica-facts-for-kids
Learn all about Antarctica and the animals that live there.

Orca Research Trust: Orca Fact Sheets for Children
www.orcaresearch.org/resources/kids-corner/fact-sheets-for-children
Read and download fact-filled booklets about killer whales.